D1543536

Alice in Wonderland

A DORLING KINDERSLEY BOOK

Produced by  Leapfrog Press Ltd

Project Editor Naia Bray-Moffatt
Art Editor Penny Lamprell
Designers Catherine Goldsmith and Adrienne Hutchinson

For Dorling Kindersley
Senior Editor Marie Greenwood
Managing Art Editor Jacquie Gulliver
Picture Research Liz Moore
Production Joanne Rooke

First published in Great Britain in 2000 by Dorling Kindersley Limited
9 Henrietta Street, London WC2E 8PS
2 4 6 8 10 9 7 5 3 1
www.dk.com

Compilation copyright © 2000 Dorling Kindersley Limited
Illustration copyright © 2000 Greg Becker
Text copyright © 2000 Jane Fior
All rights reserved. No part of this publication may be reproduced, stored in a retrieval system or
transmitted in any form or by any means, electronic, mechanical, photocopying, recording or otherwise,
without the prior written permission of the copyright owner.

ISBN 0-7513-7110-6
Colour reproduction by Bright Arts, Hong Kong
Printed in Italy by L.E.G.O.
A CIP catalogue record for this book is available from the British Library.

Acknowledgements
The publisher would like to thank the following for their kind permission to reproduce their photographs:
a=above; c=centre; b=bottom; l=left; r=right; t=top

AKG London Ltd: 48bl; Bridgeman Art Library: 7b, 18br, 29tl, 30, 33b, 36b, 45tl, br, 47tl;
ET Archive: 48cr; Mary Evans Picture Library: 11tr, 21tl, 25tr, 34br, 44tl, 46br, 47bl, 48tl,
48 background; Hulton Getty: 46cr, 48tr.

YOUNG DK CLASSICS

Alice in Wonderland

By **Lewis Carroll**
Adapted by **Jane Fior**

Illustrated by
Greg Becker

DK
Dorling Kindersley

Contents

Down the Rabbit Hole

ALICE WAS BEGINNING TO GET VERY TIRED of sitting by her sister on the bank, and of having nothing to do: once or twice she had peeped into the book her sister was reading, but it had no pictures or conversations in it. And what is the use of a book, thought Alice, without pictures or conversations?

She was wondering whether to get up and make a daisy chain when suddenly a white rabbit with pink eyes ran close by her. He took a watch out of his waistcoat pocket and muttered "Oh dear! Oh dear! I shall be late!"

Alice had never seen a rabbit with a watch or a waistcoat pocket before, and so, burning with curiosity, she ran after it and was just in time to see the Rabbit disappear down a large rabbit hole under the hedge. Without thinking twice, Alice went down after it.

The rabbit hole went straight on like a tunnel but then it suddenly dipped so that Alice found herself falling down what seemed like a very deep well. As she fell, she noticed that the sides of the well were filled with cupboards and bookshelves and maps hanging on pegs.

A white rabbit with pink eyes ran close by her.

Down, down, down. Would the fall
never come to an end? "I wonder how
many miles I have fallen?" said Alice.
"I must surely be somewhere near the centre
of the earth by now. Perhaps I shall fall right through!"

*She landed on a heap of
leaves and sticks.*

On and on she fell. Alice soon began talking to herself again.
"Dinah will miss me very much tonight," she said. (Dinah was
the cat.) "I hope they remember her saucer of milk at tea-time.
Oh, Dinah, my dear, I do wish you were down here with me."

Alice was beginning to feel quite drowsy when thump! thump!
she landed on a heap of leaves and sticks and
her fall was over. She was not a bit hurt, and
she jumped to her feet in a moment. She
looked up, but it was all dark overhead. Before
her was another long passage and the White
Rabbit was still in sight, scurrying down it.
There was not a moment to be lost. Away went
Alice like the wind, and she was just in time to
hear the Rabbit say, as it turned a corner, "Oh
my ears and whiskers, how late it's getting!"

*Pocket watches like the one
worn by the White Rabbit
were used until quite
recently. They were attached
to a chain for safe-keeping.*

Alice ran round the corner as fast as she
could but the Rabbit was no longer to be seen.
She found herself in a long, low hall, which was lit up by a row
of lamps hanging from the roof. There were doors all around
her but every one was locked. Alice wondered sadly
how she was ever to get out again.

*On top of the table she
found a golden key.*

Suddenly she came upon a little three-legged table made of solid glass and on top of it she found a golden key. Alice thought that it might belong to one of the doors, but alas! either the locks were too large or the key too small – at any rate it would not open any of them.

Then, to her surprise, she noticed a low curtain and hidden behind it a little door. She tried the golden key in the lock and to her great delight found that it fitted.

Alice opened the door and found herself looking down a small passage. It opened out onto the loveliest garden you ever saw. Alice wished she could wander about those beds of bright flowers, but she could hardly get her head through the doorway. "If only I could shut up like a telescope," said Alice. "I do believe I could if I only knew how to begin."

She went back to the table, half hoping to find a book of instructions, but instead she found a little bottle. There was a label tied to the bottle and on it were the words DRINK ME.

*The passage opened on
the loveliest garden.*

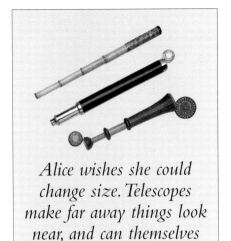

Alice wishes she could change size. Telescopes make far away things look near, and can themselves fold up to become small.

Wise little Alice did not drink the contents of the bottle straight away. She picked it up and looked at it carefully to see whether it was marked "poison", for she knew that if you drink from a bottle marked "poison" it is certain to disagree with you sooner or later.

However, this bottle was not marked "poison" and so Alice ventured to taste it and she found it very nice. It had a sort of mixed flavour of cherry tart, custard, pineapple, roast turkey, toffee, and hot buttered toast. Alice soon finished it off.

Then Alice had a very curious feeling. "Goodness," she said. "I seem to be shutting up like a telescope!"

And indeed, so she was. She was now only ten inches tall! In fact, she was just the right size to go through the little door into that lovely garden. She waited for a few minutes to see if she was going to shrink any more, but when nothing else happened she decided to go into the garden at once.

But, oh dear, when she reached the door she found that she had forgotten the key and when she went back to the table, she could not possibly reach it although she could see it quite plainly through the glass. She did her best to climb up one of the table legs but it was too slippery.

Tired out with trying, Alice sat on the floor and wept.

Alice sat on the floor and wept.

In fact she was now more than nine feet tall.

"There's no use crying like that," said Alice to herself rather sharply. "Stop it!" It was then that she saw a little glass box lying under the table and in it a small cake with the words EAT ME beautifully marked out in currants. "I will eat it," said Alice, "for if it makes me grow larger I can reach the key, and if I grow smaller I can creep under the door, so either way I'll get into the garden." She ate a little bit and then a little bit more until it was finished.

"Curiouser and curiouser," cried Alice. "I am opening out like the biggest telescope there ever was." She looked down at her feet and found that they were now so far off, they were almost out of sight.

And at that moment her head struck against the roof of the hall. In fact she was now more than nine feet tall. She seized the little key and hurried back to the garden door.

Poor Alice! She could only manage to see the garden by lying on her side. Getting through the passage was more hopeless than ever. Alice sat down and began to cry again and she was soon surrounded by a large pool of tears.

After a time she heard the pattering of feet in the distance. It was the White Rabbit returning, splendidly dressed, with a pair of white gloves in one hand and a large fan in the other. He was muttering, "Oh, the Duchess! The Duchess! She will be so angry to be kept waiting!"

Alice tried to speak to him but the Rabbit started violently, dropped the gloves and fan and scurried away into the darkness.

Alice picked up the fan. "Dear me, how strange everything is today," she said, "and yesterday things were quite normal. I wonder if I've been changed in the night?"

As she spoke, she was trying on one of the Rabbit's little kid gloves. "How can I have done that?" she exclaimed. "I must be growing small again."

Measuring herself against the table, she found that she was now only two feet high and still shrinking. The fan was the cause so Alice dropped it quickly just in time to save herself from shrinking away completely.

"What a narrow escape!" said Alice. "And now for the garden!" She ran back to the door but alas! The little door was shut again and the key was back on the table.

"Things are worse than ever," said Alice. Just then her foot slipped and splash! she was up to her chin in salty water. "Oh, I wish I had not cried so much," said Alice as she swam about.

In Victorian times, when Carroll was writing, gentlemen changed into smart clothes every evening. They wore kid gloves made from goat skin.

Just then her foot slipped and splash!

The Pool of Tears

J UST THEN, SHE HEARD SOMETHING splashing in the pool and as she swam towards it, she saw that it was a mouse.

"Excuse me, Mouse," said Alice politely. "Do you know the way out of this pool?"

The mouse stared at her but said nothing.

"Perhaps it doesn't understand English," said Alice so she tried again. "Où est ma chatte?" (This was the first sentence in her French lesson book and it meant "Where is my cat?")

The mouse leaped out of the water.

"Goodness," said Alice, "I do beg your pardon. I forgot that you don't like cats."

"Not like cats!" cried the Mouse passionately. "Would you like cats, if you were me?"

"I am sorry, Mouse," said Alice. "I won't mention them again. Are you fond of dogs, perhaps?" At this, the mouse began to swim hastily away and Alice realized that she had offended it again.

"Mouse, dear," called Alice, "do come back and we won't talk about cats or dogs."

"Let us get to the shore and then I'll tell you why I hate cats and dogs," said the Mouse.

When the Mouse heard this, it turned round and swam slowly back to her, its face pale. "Let us get to the shore and then I'll tell you why I hate cats and dogs," said the Mouse.

The pool was getting quite crowded with all the birds and animals that had fallen into it. There was a Duck and a Dodo, a Lory and an Eaglet and several other curious creatures. Alice led the way, and the whole party swam to the shore where they gathered on the bank, all dripping wet and cross and uncomfortable.

The first question was how to get dry again. They had a consultation about this and after a few minutes Alice found herself talking to them as if she had known them all her life. Indeed, she had quite a long argument with the Lory.

At last, the Mouse, who seemed to be a person of some authority among them, called out, "Sit down all of you and listen. I'll soon make you dry." So they all sat down in a circle with the Mouse in the middle. Alice shivered. She felt sure she would catch a bad cold if she did not get dry very soon.

The Mouse began to lecture them all about history. The animals fidgeted but the Mouse went on and on. Finally he turned to Alice and said, "How are you now, my dear?"

"As wet as ever," complained Alice.

"In that case," said the Dodo, rising to its feet, "I suggest that we adopt more energetic measures."

"Speak English!" said the Eaglet. "I don't know the meaning of half those long words and what's more, I don't believe you do either!" At that, some of the animals began to giggle.

"I suggest," said the Dodo in an offended tone, "that the best thing to get us dry would be a Caucus-race."

"What is a Caucus-race?" asked Alice.

"The best way to explain it is to do it," said the Dodo and he proceeded to mark out a race course in a sort of circle. He then placed all the animals along it, here and there. There was no "One, two, three and away!" They all began running when they liked and left off when they liked, so it was hard to know when the race was over. However, when they had all been running for half an hour or so, and were quite dry again, the Dodo suddenly called out, "The race is over!"

Everyone crowded round and asked, "Who has won?" The Dodo thought hard and finally he announced, "Everybody has won and all must have prizes."

"But who is to give out the prizes?" a chorus of voices asked.

"Why, she is, of course," said the Dodo, pointing at Alice. Alice had no idea what to do, and in despair she put her hand in her pocket and pulled out a box of sweets (luckily the water had not got into them) and handed them round as prizes. There was exactly one each.

When they had all finished, they sat down in a circle again and begged the Mouse to tell them something more.

"You promised to tell me why you hate C... and D...," whispered Alice, afraid that the Mouse might be offended again.

"Mine is a long and sad tale," said the Mouse, sighing.

"It is a long tail, certainly," said Alice, gazing at the Mouse's tail, "but why do you call it sad?" She kept on puzzling about this while the Mouse was speaking until it said severely, "You are not attending!" and got up to leave.

"Please come back and finish your story," begged Alice.

"Yes, do," said the others.

But the Mouse only shook its head impatiently and walked away.

"I wish I had Dinah here," remarked Alice. "She'd soon fetch the Mouse back."

"And who is Dinah?" asked the Lory.

Alice replied, "Dinah's our cat. She's so good at catching mice! And I wish you could see her after the birds! Why, she'll eat a little bird as soon as look at it!"

This speech caused a sensation. Some of the birds hurried off at once, and a Canary called out to its children, "Come away, my dears! It's high time you were all in bed."

One after another they presented their excuses and left, and soon Alice found herself alone once more.

———

"What is a Caucus-race?"
asked Alice.

There was a pattering of footsteps and, looking up, Alice saw the White Rabbit trotting back, muttering to itself, "The Duchess! Oh my paws and whiskers! Where can I have dropped them?" Alice guessed it was looking for the gloves and fan and she began to hunt for them but they were nowhere to be seen.

Very soon, the Rabbit noticed Alice and it called out in an angry tone, "Mary Ann, what are you doing here? Run home this minute." Alice was so frightened that she ran off at once.

"He took me for his housemaid," she said. "How strange to be running errands for a rabbit!"

As she said this, she noticed a little house, and on the door a brass plate with W. Rabbit engraved upon it. She went in without knocking and hurried straight upstairs to look for the fan and gloves.

She found herself in a tidy little room with a table in the window and on it, as she hoped, a fan and two or three pairs of tiny white gloves. On the dressing table stood a little bottle. It had no label this time but she uncorked it and put it to her lips. "I do hope this will make me grow large again for really I am quite tired of being so small."

It did, and much sooner than she expected, for before she had drunk half the bottle she found her head pressing against the ceiling. Soon she had to kneel on the floor and still she kept growing. To make more room for herself, she put one arm out of the window and one foot up the chimney. "Now I can do no more," said Alice. "What will become of me?"

Luckily for Alice, the little bottle had now had its full effect but she felt very uncomfortable and wondered how she would ever get out of the room.

It was then that she heard someone rattling at the door and she heard the Rabbit say, "Mary Ann, fetch my gloves this minute." Then she heard someone remark on her arm sticking out of the window and her foot sticking out of the chimney. And this was followed by a shower of little pebbles rattling at the window. Some of them even hit her face.

Alice then noticed that the pebbles were turning into little cakes as they lay on the floor and suddenly she had an idea.

"If I eat one of these cakes," she thought, "it's sure to make some change in my size, and since it can't make me larger, it must make me smaller, I suppose."

So she swallowed one of the cakes and was delighted to find that she immediately began to shrink.

"Mary Ann, fetch my gloves this minute."

Soon Alice was small enough to get out of the house. There was a crowd of angry animals and birds by the door but Alice ran off as fast as she could and found herself safe in a thick wood.

"The first thing I've got to do," said Alice, "is to grow to my right size again, and the second is to find my way into that lovely garden."

This was an excellent plan but unfortunately she had not the slightest idea how to go about it. "I suppose I ought to eat or drink something but the question is 'What?'"

Alice looked around her but she could see nothing suitable. There was a large mushroom growing close by, so Alice decided to look and see what was on top of it.

She stretched up on tiptoe and her eyes immediately met those of a large blue caterpillar. It was sitting on top, quietly smoking a long hookah.

"Who are you?" asked the Caterpillar.

Alice replied shyly, "I hardly know, Sir. I know who I was when I got up this morning but I have changed several times since then."

"What do you mean by that?" said the Caterpillar sternly.

"I'm afraid I can't put it more clearly," said Alice. "Perhaps you could tell me who you are instead."

"Why?" said the Caterpillar. It seemed to be in a bad temper so Alice turned to go.

"Come back!" the Caterpillar called after her. "So you think you've changed do you?"

"I am afraid so," said Alice. "I don't seem able to stay the same size for more than ten minutes."

Hookah (tobacco) pipes from the Middle East were considered fashionable and exotic. This Caterpillar is an exotic creature.

"Who are you?" asked the Caterpillar.

"What size do you want to be?" asked the Caterpillar.

"Well, I should like to be a little larger, Sir. Three inches is such a tiny size to be."

"It is a very good size," said the Caterpillar crossly.

"But I'm not used to it," said Alice.

The Caterpillar yawned. "One side of this mushroom will make you grow taller," it said. "The other will make you grow shorter." And it slithered down and crawled away through the grass.

"But which side is which?" wondered Alice. She broke off a piece from one and a piece from the other and nibbled a little right-hand bit just to see. The next moment she felt herself shrinking rapidly so she took a bite of the left-hand piece, only to shoot up until she was taller than the trees. She continued to eat, first one piece then another until finally she found herself her usual size.

It was then that she saw the little house. It was about four feet high. "I wonder who lives there?" said Alice. "I need to be smaller if I am to visit," so she nibbled on the mushroom till she had brought herself down to nine inches tall.

Pig and Pepper

JUST THEN, a footman came running out of the wood and rapped loudly on the door. He was in livery and reminded Alice of a fish. The door was opened by another footman who looked just like a frog. The Fish-Footman produced a large letter and handed it over. "For the Duchess," he said, "an invitation from the Queen to play croquet."

The two footmen then bowed low and their curls got tangled together. Alice laughed so much she had to run back into the wood and when she next peeped out, the Fish-Footman was gone and the other was sitting on the ground.

Alice went up to the door. "There's no use in knocking," said the Footman. "They are making far too much noise inside." And certainly there was a most extraordinary noise going on, a constant howling and sneezing, and every now and then a great crash as if a dish had been broken.

At this moment, the door of the house opened and a plate came skimming out. Alice stepped inside and found herself in a large kitchen which was full of smoke. The Duchess was sitting on a stool, holding a baby; the cook was leaning over the fire, stirring a large cauldron of soup.

The cook took the soup off the stove.

The phrase "Grin like a Cheshire cat" was a popular one, but no one quite knows where it comes from. It might be because Cheshire cheeses used to be moulded in the shape of a grinning cat.

Alice sneezed. "There's certainly too much pepper in that soup," she said. The Duchess was sneezing too and so was the baby. The only two creatures not sneezing were the cook and a large cat which was lying on the hearth and grinning from ear to ear.

"Please would you tell me," said Alice politely, "why your cat grins like that?"

"It's a Cheshire cat," said the Duchess. "That's why."

"I didn't know that cats could grin."

"They all can, and most of 'em do," said the Duchess.

As Alice wondered what to say next, the cook took the soup off the stove and began to throw everything in reach at the Duchess and the baby – saucepans, plates and dishes.

"Oh, please mind what you are doing!" cried Alice.

"If everybody minded their own business," said the Duchess, "the world would go round a good deal faster."

"Which would not be an advantage," said Alice.

"Oh, don't bother me!" said the Duchess. And she began to sing a sort of lullaby to the baby:

Speak roughly to your little boy,
And beat him when he sneezes:
He only does it to annoy,
Because he knows it teases.

"Here, you take it a for a bit," said the Duchess, flinging the howling baby at Alice. "I must go and get ready to play croquet with the Queen."

Alice caught the baby with some difficulty. It was a queer-shaped little creature and lay snorting and squirming in her arms.

"If I don't take this child away with me," said Alice, "they're sure to kill it in a day or two. Don't grunt," she said to the baby, "it's not polite."

The baby grunted again and Alice looked at it anxiously. It had a very turned-up nose, much more like a snout than a real nose; also its eyes were getting extremely small for a baby. "Perhaps it was only sobbing," said Alice and she looked into its eyes to see if there were any tears.

No, there were no tears. "If you're going to turn into a pig, my dear," said Alice seriously, "I'll have nothing more to do with you!"

The baby grunted again. Alice looked down in some alarm. This time there could be no mistake. The creature was indeed a pig. She set it down and watched it trot off into the wood.

"If it had grown up," thought Alice, "it would have made a dreadfully ugly child but it makes rather a handsome pig!" As she was thinking this, she noticed the Cheshire Cat sitting in a tree.

The Cat grinned. It looked good-natured though Alice noticed that it had very long claws and a great many teeth.

"Cheshire-Puss," said Alice, "which way ought I to go from here?"

"It depends on where you want to get to," said the Cat.

"I don't much care where, as long as I get somewhere," said Alice.

"You're sure to do that if only you walk long enough," said the Cat.

The creature was indeed a pig.

"What sort of people live about here?" asked Alice.

"In that direction lives a Hatter, and in that direction lives a March Hare. Visit either if you like. They're both mad."

"But I don't want to go among mad people."

"Oh, you can't help that. We're all mad here," said the Cat. "By the way, do you play croquet with the Queen today?"

"I should like to very much," said Alice, "but I haven't been invited."

"You'll see me there," said the Cat, and vanished.

Alice waited a little to see if it would reappear but it did not, so Alice started to walk in the direction of the March Hare. When she found its house, she saw that the chimneys were shaped like ears and the roof was thatched with fur. Alice felt anxious. "Suppose it should be raving mad?" she said. "I almost wish I'd gone to see the Hatter instead!"

The Cat grinned.

A Mad Tea Party

THERE WAS A TABLE SET OUT in front of the house, and the March Hare and the Hatter were having tea at it. A Dormouse was sitting between them, fast asleep, and the other two were using it as a cushion.

The table was a large one but the three were all crowded together at one corner. "No room! No room!" they cried out when they saw Alice. "There's plenty of room," said Alice and she sat down in a large armchair at one end of the table.

"Have some wine," said the March Hare.

"I don't see any wine," said Alice.

"There isn't any," said the March Hare.

The Hatter opened his eyes wide. "Why is a raven like a writing desk?" he said. They sat silently and Alice thought over all she knew about ravens and writing desks, which wasn't much.

The Hatter took out his watch.

"What a funny watch," said Alice. "It tells the day of the month but doesn't tell what o'clock it is."

"Why should it?" muttered the Hatter.

"Have you guessed the answer to the riddle yet?" asked the Hatter.

"No, I give up," Alice replied. "What's the answer?"

"I haven't the slightest idea."

Alice sighed. "Why do you waste time asking riddles that have no answers?"

"If you knew Time as well as I do," said the Hatter, "you wouldn't talk about wasting it. We fell out last March at the great concert given by the Queen of Hearts. I had to sing, 'Twinkle, twinkle little bat! How I wonder what you're at.' I had hardly started, when the Queen bawled out, 'He's murdering the time! Off with his head!' Since then, Time won't do a thing I ask. It's always six o'clock now."

The phrase "mad as a hatter" comes from a time when hatters (hat makers) used mercury to make hats. Exposure to mercury caused rather odd behaviour which some thought was a kind of madness.

"Is that why there are so many tea-things put out?" asked Alice.

"Yes, it's always tea-time here," said the Hatter.

"Suppose we change the subject," said the March Hare. "Let's ask the Dormouse to tell us a story."

"Once upon a time there were three little sisters," the Dormouse began, "and their names were Elsie, Lacie and Tillie, and they lived at the bottom of a well –"

"What did they live on?" asked Alice

"They lived on treacle," said the Dormouse.

"And why did they live at the bottom of a well?"

The Dormouse thought for a minute or two. "It was a treacle well," he said.

"There's no such thing," said Alice.

"Sh! Sh!" said the Hatter and the March Hare, so Alice promised not to interrupt. The Dormouse continued, "And these three little sisters, they were learning to draw."

"What did they draw?" said Alice, forgetting her promise.

"Treacle," said the Dormouse.

"I want a clean cup," said the Hatter suddenly. "Let's all move one place on."

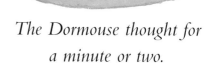

The Dormouse thought for a minute or two.

He changed places as he spoke. They all followed him and Alice found herself in the March Hare's seat. Not wishing to offend the Dormouse again, she enquired cautiously, "Where did they draw the treacle from?"

"You can draw water out of a water well," said the Hatter, "so I should think you could draw treacle out of a treacle well – eh, stupid?"

"But they were in the well already," said Alice.

"Of course they were," said the Dormouse, "well in."

This answer confused Alice but she let the Dormouse continue without interrupting. "They were learning to draw," said the Dormouse yawning and rubbing its eyes, "and they drew all manner of things beginning with M, such as mouse-traps, and the moon, and memory, and muchness. You know how you say things are "much of a muchness". Did you ever see such a thing as a drawing of a muchness?"

"Well," said Alice, "I don't think –"

"Then you shouldn't talk!" said the Hatter.

This piece of rudeness was more than Alice could bear. She got up and walked off. The Dormouse fell asleep at once and the other two took no notice of her. She looked back to see the Hatter and the March Hare trying to put the Dormouse into the teapot.

"I'll never go there again," said Alice. "That was the stupidest tea party I ever was at in all my life!"

Just then, she noticed a tree with a door in its trunk. "That's curious!" she said, "but then, everything's curious today. I think I'll go in at once." And in she went.

Once again she found herself in the long hall. "I'll manage better this time," she said, and taking the little golden key, she unlocked the door that led to the garden. She nibbled at the mushroom (she had kept a bit in her pocket) till she was the right size and walked down the passage. At last she was in the beautiful garden!

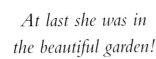

At last she was in the beautiful garden!

The Queen's Croquet Ground

A LARGE ROSE TREE stood at the entrance to the garden. The roses were white but there were three gardeners busily painting them red. As Alice went closer, she heard one say, "Look out now, Five! Don't splash the paint like that."

"I couldn't help it," said Five. "Seven jogged my elbow."

"That's right, Five! Always blame somebody else!"

"You can't talk," said Five. "I heard the Queen say yesterday that you deserved to be beheaded."

"What for?" said the one who had spoken first.

"None of your business, Two," said Seven. He flung down his brush angrily but caught sight of Alice and checked himself. He immediately started to bow and then the other two bowed as well.

"Would you tell me please, why you are painting those roses?" asked Alice a little timidly.

Two replied in a low voice, "You see, Miss, this ought to have been a red rose tree but we put a white one in by mistake. If the Queen finds out, she'll have our heads cut off for sure."

At this, Five called out anxiously, "The Queen! The Queen!" and the three gardeners immediately threw themselves face down on the ground.

Then came the King and Queen of Hearts!

Victorian children spent many hours playing cards. They would have enjoyed reading about the cards coming to life in the story.

Alice looked round to see ten soldiers carrying clubs. Like the gardeners, they were quite flat, with their hands and feet at the corners. They were followed by ten courtiers ornamented with diamonds. After these came the royal children, all ten of them jumping merrily and ornamented with hearts. Next came the guests, mostly Kings and Queens and amongst them Alice recognized the White Rabbit. Then came the King and Queen of Hearts!

When the procession reached Alice, they all stopped and looked at her and the Queen said severely, "Who is this?"

"My name is Alice, so please your Majesty," said Alice very politely, reminding herself not to be frightened for they were only a pack of cards after all.

"And who are these?" said the Queen, pointing to the gardeners. As they were lying on their faces, the pattern on their backs was the same as the rest of the pack, so the Queen could not see if they were gardeners or indeed her own children.

"How should I know?" said Alice.

The Queen turned crimson with fury and began to scream, "Off with her head! Off with her head!"

"Nonsense!" said Alice very loudly.

The Queen was silent. The King laid his hand on her arm and said, "Consider, my dear; she is only a child."

The Queen stared down at the gardeners. She turned to the Knave of Hearts. "Turn them over," she said. And so, very carefully, with one foot, the Knave turned the gardeners over. They immediately jumped up and began to bow.

"Stop that!" shrieked the Queen. "You're making me giddy." She turned to the rose-tree. "What's this?"

"May it please your Majesty," began Two. "We —"

"Off with their heads!" said the Queen. The soldiers approached and the gardeners ran to Alice for protection.

"You shan't be beheaded," said Alice and she put them into a flowerpot.

"Are their heads off?" shouted the Queen.

"Their heads are gone," the soldiers shouted in reply.

The Queen turned to Alice. "Do you play croquet?" she shrieked.

"Yes!" shouted Alice.

"Come on then," said the Queen, and Alice joined the procession.

"It's — it's a very fine day," said a timid voice at her side. She was walking by the White Rabbit.

"Very," said Alice. "Where's the Duchess?"

"Sh!" said the Rabbit. He whispered in her ear, "She's under sentence of execution."

"Get to your places!" shouted the Queen, and everybody began to run about in all directions.

Croquet was a popular outdoor game with wealthy people who owned large gardens. Lewis Carroll enjoyed the game.

*Alice found it very difficult
to manage her flamingo.*

Once they were sorted out,
the game began.

Alice thought that she had never seen such a curious
croquet lawn in all her life. It was all ridges and furrows, the croquet
balls were live hedgehogs, the mallets live flamingoes and the soldiers
had to double themselves up to make the arches. Alice found it very
difficult to manage her flamingo. As soon as she had it in position, it
would twist itself round and stare at her with such a puzzled
expression that Alice had to laugh. All the while, the other players
were quarrelling and fighting and the Queen was marching around,
angrily shouting "Off with their heads!"

Alice felt nervous. Although she had not yet quarrelled with the
Queen, it could happen at any moment. Alice decided she had to
find a way to escape.

She found the Cat surrounded by a large crowd.

As Alice looked about her, she noticed a curious apparition in the air. Watching carefully, she made out a grin. "It's the Cheshire Cat!" cried Alice. "Now I shall have somebody to talk to!"

"How are you getting on?" asked the Cat as soon as there was enough mouth for it to speak with. "How do you like the Queen?"

"Not at all," said Alice. "She's so extremely −" Just then she noticed the Queen standing close behind her so she went on, "− likely to win the game."

"Who are you talking to?" said the King.

"It's a friend of mine, a Cheshire Cat," said Alice.

"I don't like the look of it," said the King. "It must be removed."

The Queen immediately said, "Off with its head!" so the King hurried away to fetch the executioner.

The croquet game was in great confusion. Alice went in search of her flamingo but by the time she had caught it, the game was over so she tucked it under her arm and went back to look for the Cheshire Cat.

She found it surrounded by a large crowd. There was a dispute going on between the Queen and the executioner who was pointing out that you couldn't cut off a head unless there was a body to cut it off from. The Queen said angrily that if something wasn't done about the Cheshire Cat at once, she'd have everybody executed.

She was about to begin a long conversation.

"The Cat belongs to the Duchess," said Alice. "Perhaps you should talk to her about it."

"She's in prison," said the Queen. She turned to the executioner. "Fetch her here." As soon as he had gone, the Cat's head began to fade and by the time the executioner returned with the Duchess, it had disappeared completely.

"You can't think how glad I am to see you again, you dear old thing," said the Duchess, coming over to Alice. She was about to begin a long conversation when the Queen came and stood in front of them, frowning like a thunderstorm.

"A fine day, your Majesty," began the Duchess.

The Queen stamped her foot. "Now, I give you fair warning," she shouted. "Either you or your head must be off. Take your choice!"

The Duchess took her choice and was off in a moment.

"Have you seen the Mock Turtle yet?" asked the Queen.

"No," said Alice. "I don't even know what a mock turtle is."

"It's the thing Mock Turtle soup is made from," said the Queen. "Come with me and hear his story."

Mock turtle soup was a favourite Victorian dish. It is a pretend green turtle soup, made from veal, which was much cheaper than real turtle.

The Mock Turtle's Story

As THEY WALKED, they came upon a Gryphon, lying fast asleep in the sun. (If you don't know what a Gryphon is, look at the picture.) "Up, lazy thing," said the Queen, "and take this young lady to see the Mock Turtle. I must go back and order some more executions." And she walked off, leaving Alice alone with the Gryphon.

Alice did not like the look of the creature but she sat down and waited for it to wake up. Her patience was rewarded. The Gryphon yawned and rubbed its eyes. It watched until the Queen was out of sight, then it said, "What fun!"

"What is the fun?" said Alice.

"Why, she is!" said the Gryphon. "They never execute anybody. She just imagines it. Come on!"

As they walked, they came upon a Gryphon.

Everybody says 'come on' here, thought Alice. I was never so ordered about in all my life.

They had not gone far when they found the Mock Turtle, sitting sad and lonely on a little ledge of rock. He was sighing as if his heart would break.

"What is his sorrow?" asked Alice.

"He's got no sorrow," replied the Gryphon. "He's just imagining it. Come on!" He turned to the Turtle, "This young lady wants to know your history."

"I'll tell it to her then," said the Mock Turtle. "Sit down and don't say a word until I've finished. Do you know that once I was a real turtle?"

These words were followed by a

The Gryphon, or griffin, is a mythical beast with the head and wings of an eagle and the lower body of a lion.

long silence broken only by the Mock Turtle's sobs. Then it continued, "When we were little, we went to school in the sea. We had the best of educations. In fact, we went to school every day –"

"This young lady," said the Gryphon, "wants to know your history."

"I've been to a day school too," interrupted Alice.

"With extras?" asked the Mock Turtle, a little anxiously.

"Yes," said Alice, "we learned French and music."

"And washing?" said the Mock Turtle.

"Certainly not!" said Alice indignantly.

"Ah," said the Mock Turtle with relief. "Then yours wasn't a really good school. At ours, we had French and music and washing, extra."

"You couldn't have needed it much," said Alice, "living at the bottom of the sea."

"I only took the regular lessons," said the Mock Turtle. "Reeling and Writhing, and Drawling and Stretching and Fainting in Coils."

"That's quite enough about lessons," interrupted the Gryphon. "Tell her about the games."

The Mock Turtle sighed deeply. "You may not have lived much under the sea," he said, starting to sob once again, "so perhaps you have never met a lobster – (Alice began to say "I once tasted –" but she stopped herself in time and said "No, never,") – so you can have no idea what a delightful thing a Lobster-Quadrille is."

"No, indeed," said Alice. "What sort of dance is that?"

*"It must be a very
pretty dance."*

"You form a line along
the seashore," said the Gryphon.

"Two lines," cried the Mock Turtle. "Seals,
turtles, salmon and so on, and then when you've cleared the jelly-fish
out of the way, you advance twice –"

"Each with a lobster as a partner!" cried the Gryphon.

"Set to your partners and change lobsters," said the Mock Turtle.
"Then throw the lobsters as far out to sea as you can –"

"And swim after them!" screamed the Gryphon.

"Turn a somersault in the sea!" cried the Mock Turtle, capering
wildly about.

"Change lobsters!" yelled the Gryphon.

"Back to land again and that's the end of the first part!" said the
Mock Turtle.

"It must be a very pretty dance,"
said Alice timidly.

"Let's try it," said the Mock Turtle.
"We can do it without lobsters."

"You sing," said the Gryphon. "I've
forgotten the words."

They began to dance solemnly
round Alice while the Mock
Turtle sang:

*Many children would have
been taught to dance the
quadrille. It was a
fashionable, but difficult,
ballroom dance.*

"Will you walk a little faster?"said a whiting to a snail,
"There's a porpoise close behind us, and he's treading on my tail.
See how eagerly the lobsters and the turtles all advance!
They are waiting on the shingle – will you come and join the dance?
Will you, won't you, will you, won't you, will you join the dance?
Will you, won't you, will you, won't you, won't you join the dance?"

"Thank you," said Alice. "That was most interesting, though if I'd been the whiting, I'd have said to the porpoise, 'Keep back, please! We don't want *you* with us!'"

"They were obliged to have him with them," the Mock Turtle said. "No wise fish would go anywhere without a porpoise."

"Don't you mean purpose?" said Alice.

"I mean what I say. Now you tell us something."

"I could tell you my adventures, beginning from this morning," said Alice. She began to tell them everything that had happened to her and the two creatures sat very close to her, one on each side, with their eyes and mouths very wide open.

"Shall we dance the Lobster-Quadrille again?" asked the Gryphon suddenly, "or would you like the Mock Turtle to sing you another song?"

"Oh, a song please," said Alice.

"Sing her 'Turtle Soup'," said the Gryphon. The Mock Turtle sighed deeply, and began:

Beautiful Soup, so rich and green,
Waiting in a hot tureen!

His words were interrupted by a cry of "The trial's beginning!"

"What trial?" said Alice.

"Come on," cried the Gryphon, seizing her hand and running without waiting for the end of the song.

"Come on," cried the Gryphon.

"Herald, read the accusation!" said the King.

Who Stole the Tarts?

THE KING AND QUEEN OF HEARTS were seated on their thrones. The Knave was standing in front of them, in chains, and near the King was the White Rabbit, with a trumpet in one hand and a scroll of parchment in the other. In the middle of the court was a table with a large dish of tarts upon it. They looked very good.

Alice had never been in a court before but she had read about them in books. The King was the judge and he wore his crown over his wig. And twelve creatures, all busily writing, were the jurors.

"Herald, read the accusation!" said the King.

The White Rabbit blew three blasts on his trumpet and, unrolling the parchment, read out:

> *The Queen of Hearts, she made some tarts,*
> *All on a summer day;*
> *The Knave of Hearts, he stole those tarts,*
> *And took them quite away!*

"Consider your verdict," the King said to the jury.

"Not yet," the Rabbit hastily interrupted. "There's more to come."

"Call the first witness," said the King, and again the Rabbit blew three blasts on his trumpet.

The first witness was the Hatter. He came in with a teacup in one hand and a piece of bread-and-butter in the other. He was followed by the March Hare and the Dormouse.

"Take off your hat," the King said to the Hatter.

"It isn't mine," said the Hatter.

"Stolen!" the King exclaimed.

"I keep them to sell," the Hatter explained. "I've none of my own. I'm a hatter."

Here the Queen put on her spectacles and began staring hard at the Hatter who turned pale, and fidgeted.

"Give your evidence and don't be nervous or I'll have you executed on the spot," said the King.

The Hatter shifted uneasily from one foot to the other and in his confusion bit a large piece out of his teacup instead of the bread-and-butter.

"I'm a poor man, your Majesty," the Hatter began, in a trembling voice, "and I hadn't begun my tea, and what with the bread-and-butter getting so thin, and the twinkling of the tea –"

"The twinkling of what?" said the King.

"It began with the tea," the Hatter replied.

"Of course twinkling begins with a T," said the King sharply. "Do you take me for a dunce?"

Just at this moment Alice felt a very curious sensation. She realized that she was growing larger again!

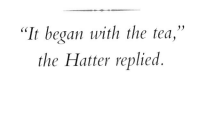

*"It began with the tea,"
the Hatter replied.*

Alice's Evidence

THE KING SAID, "CALL THE NEXT WITNESS." The next witness was the Duchess's cook. She carried the pepperbox in her hand and at once the people near the door began to sneeze.

"What are tarts made of?" asked the King in a deep voice.

"Pepper mostly," said the cook.

"Treacle," said a sleepy voice beside her.

"Behead that dormouse," shrieked the Queen.

For some minutes the court was in confusion and by the time they had settled down again, the cook had disappeared.

"Never mind," said the King. "Call the next witness."

Alice was very curious to see who the next witness would be. Imagine her surprise when the White Rabbit read out, at the top of his voice, "Alice!"

"Here!" cried Alice, quite forgetting how large she had grown. She jumped up in such a hurry that she tipped over the jury box, upsetting all the jurymen on to the heads of the crowd below.

"Oh, I do beg your pardon!" she exclaimed, picking them up.

"What do you know about this business?" the King said to Alice.

"Nothing," said Alice.

"That's very important," said the King, writing busily in his notebook. He then read out, "Rule Forty-two. All persons more than a mile high to leave the court."

"I'm not a mile high," said Alice.

"You are," said the King.

"Well, I won't go," said Alice. "It's not a regular rule; you invented it just now." The King turned pale and shut his notebook hastily. "Consider your verdict," he said to the jury.

"There's more evidence to come," said the White Rabbit.

"We have this letter from the prisoner to – to somebody."

"Is it in the prisoner's handwriting?" asked one of the jurymen.

"No," said the White Rabbit.

"Please your Majesty," said the Knave. "I didn't write it and I didn't sign it."

"That proves his guilt," said the Queen, "so off with −"

"It doesn't prove anything of the sort!" said Alice.

The King looked round angrily. "Let the jury consider their verdict," he said.

"No," said the Queen. "Sentence first, verdict afterwards."

"Stuff and nonsense!" said Alice loudly.

"Hold your tongue," said the Queen, turning purple.

"I won't!" said Alice.

"Off with her head!" the Queen shouted. Nobody moved.

"Who cares for *you?*" Alice said (she had grown to her full size by this time). "You're nothing but a pack of cards!"

"I'm not a mile high,"
said Alice.

At this, the whole pack rose up into the air, and came flying down upon her. Alice gave a little scream, half of fright and half of anger, and tried to beat them off, and then, suddenly, she found herself back lying on the bank, with her head in the lap of her sister, who was gently brushing away some dead leaves that had fluttered down from the trees upon her face.

"Wake up, Alice dear," said her sister. "Why, what a long sleep you've had."

"Oh, I've had such a curious dream," said Alice. And she told her sister, as well as she could remember them, all these strange adventures of hers that you have just been reading about.

Alice thought what a wonderful dream it had been.

"It certainly was a curious dream," said her sister, "but now run in for your tea; it's getting late."

So Alice got up and ran back to the house, thinking while she ran, as well she might, what a wonderful dream it had been.

Alice's Land

ALICE WAS A REAL little girl. She lived with her family in an Oxford college called Christ Church in the mid-19th century. When she was four, Alice met Lewis Carroll, who taught mathematics. He loved to tell stories, and he wrote one just for Alice.

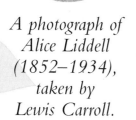

A photograph of Alice Liddell (1852–1934), taken by Lewis Carroll.

Alice loved white rabbits. In Carroll's story, it is a rabbit that leads Alice into Wonderland.

♠ ALICE'S HOME

Alice's father was the Dean of Christ Church (head of the cathedral and the college). Her family lived in the Deanery.

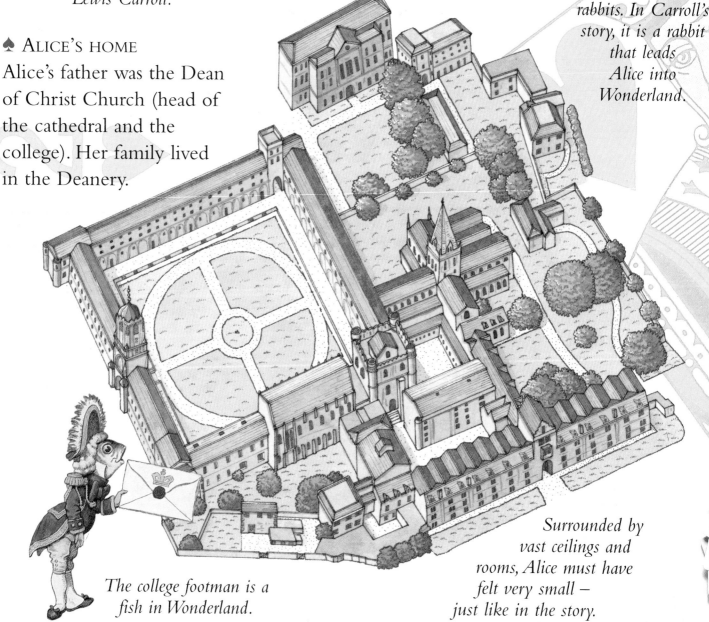

The college footman is a fish in Wonderland.

Surrounded by vast ceilings and rooms, Alice must have felt very small — just like in the story.

The Cheshire Cat.

♠ A MAD LOT
Lewis Carroll usually had his tea in Christ Church hall, where he was surrounded by

The Mad Hatter.

eccentric dons (teachers at the university). Some people believed that _Alice in Wonderland_ was a gentle satire on the odd characters he found himself among!

♠ DISAPPEARING CAT
Alice's sister, Lorina, had a cat called Dinah. Alice liked drawing Dinah as she sat in the tree. But Dinah liked to tease her by disappearing among the leaves. In _Wonderland_, Lewis Carroll makes his Cheshire Cat the ultimate disappearing cat.

"Off with her head!" said the Queen.

♠ INSIDE AND OUT
Alice, like many children at the time, played cards if it was too rainy to go outside. But croquet was her favourite outdoor game. Carroll describes a croquet game played by a pack of cards, led by the fearsome Queen of Hearts.

How very strange, thought Alice.

The Meaning of Alice

WHEN ALICE FALLS down the rabbit hole into Wonderland she enters a dream-like world full of amazing talking animals, peculiar people, changing sizes, and nonsense rhymes. Since the book was first published in 1865, people have tried to work out what *Alice in Wonderland* is all about and what the story really means.

An illustration by Tenniel.

♠ A FUN STORY
Lewis Carroll said that the story was meant only to entertain his friend Alice and her sisters while out rowing on the river and picnicking. To make the story more interesting to his listeners, he included lots of everyday things they would recognise.

A boating party.

♠ DARWIN AND ALICE
A few years before Lewis Carroll made up his story, a scientist called Charles Darwin began a famous debate about how humans originally developed from apes. Lewis Carroll may have been influenced by Darwin's ideas when he invented animals who have human characteristics.

Charles Darwin 1809–1802.

Alice (right) with her sisters.

♠ ALICE AND FRIENDS

Lewis Carroll included all the members of the Oxford boat trip in his story. The character of Alice was based on Alice Liddell.

Lorina, Alice's eldest sister, was represented by the lory, the Australian parrot.

♠ DODO

Lewis Carroll included himself as the Dodo because of his stammer. When he tried to say his real name, Charles Dodgson, it often came out as Charles Do-do-dodgson. He took the children to see the remains of a dodo at the University Museum in Oxford.

Another friend on the boat trip was the Reverend Duckworth, who was included as the duck.

Lewis Carroll's study in Oxford.

♠ CHILDREN'S GAMES

Lewis Carroll loved children and games. His study in Oxford was filled with toys that children would enjoy. One of Alice's favourite toys was a mechanical moving rabbit that lived in a cabbage. The story begins with Alice following a white rabbit.

Lewis Carroll

LEWIS CARROLL'S REAL NAME was Charles Lutwidge Dodgson. He was born in Cheshire in 1832 and was the eldest of ten brothers and sisters. As a child, he liked nothing better than to entertain his siblings with games and stories.

Lewis Carroll 1832–1898

Lewis Carroll was one of a large family.

♠ MEETING ALICE

Lewis Carroll went to Oxford to study Mathematics and Theology. At Oxford, he became friends with Alice Liddell, her two sisters and her brother. Punting on the river one day, Carroll made up a story for Alice, which we know as *Alice in Wonderland.*

Alice Liddell

♠ ALICE AND AFTER

Tenniel's illustration of Alice

Lewis Carroll paid an artist, John Tenniel, to illustrate *Alice in Wonderland,* and a publisher, Macmillan, to print his book in 1865. At first, a lot of critics did not like it. But children loved it and soon many more copies were printed. The book became so successful that Lewis Carroll wrote another *Alice* book - *Alice Through the Looking-Glass and What Alice Found There.* Both books have remained popular worldwide.